MW01680617

ANIMAL
SURVIVORS
of the
WETLANDS

ANIMAL SURVIVORS of the WETLANDS

Barbara A. Somervill

SCHOLASTIC INC.
New York Toronto London Auckland Sydney
Mexico City New Delhi Hong Kong Danbury, Connecticut

For Lilly, who makes every day an adventure

Note to readers: Definitions for words in **bold** can be found in the Glossary at the back of this book.

Photographs © 2007: Animals Animals: 38 (Steven David Miller), 21 (Patti Murray), 34 top (Maresa Pryor), 2 (Herb Segars); AP/Wide World Photos: 43 (Peter Cosgrove), 26 (George Gentry/US Fish and Wildlife Service); Corbis Images: 6 (Kevin Fleming), 44 (Amos Nachoum), 13 (Galen Rowell), 42 (Kevin Schafer), 24 (Nik Wheeler); Dembinsky Photo Assoc./Scott T. Smith: 8; Earth Scenes/Doug Wechsler: 10, 11; Minden Pictures/Frans Lanting: cover, 1, 3; Nature Picture Library Ltd./David Pike: 46; Peter Arnold Inc.: 5 top, 52 (Steve Kaufman), 16 (A.&J. Visage), 28 (Kathy Watkins); Photo Researchers, NY: back cover, 30 (D.P. Burnside), 22 (Alan D. Carey), 32 (Tim Davis), 40 (Douglas Faulkner), 49 (William Gause), 45 (Jeff Greenberg), 34 bottom (Dan Guravich), 36 (William H. Mullins), 5 left, 19 (Mark D. Phillips); University of Florida/IFAS/Tom Wright: 14.

The photograph on the cover shows an alligator with its mouth open. The photograph opposite the title page shows two manatees. The photograph on the back cover shows a brown pelican.

© 2004 by Scholastic Inc.
All rights reserved.
ISBN: 0-439-88145-5

First printing, September, 2006

Contents

A female alligator remains near her nest for the full 65 days it takes for the eggs to hatch. When she hears the hatchlings emit high-pitched croaks, she digs to help them escape the nest.

Chapter One

The Wetlands

A female alligator guards her nest deep in Georgia's Okefenokee Swamp. Her eggs should hatch any day now, and dozens of alligator **hatchlings** will scurry into the swamp's tea-colored water. Nearby, a raccoon steals an egg from a heron's nest while the hungry mother heron is away, looking for fish.

Huge mangrove trees send roots deep through swamp water into the solid earth below. Bullfrogs croak and crickets chirp from reeds and bulrushes lining a still pond. This wetland is home to shy deer and skittish rabbits, playful otters and slithering snakes. In all, the swamp is home to hundreds of mammals, thousands of birds, and millions of insects.

Defining the Wetlands

Wetlands are lands covered with salt, fresh, or **brackish water**. In North America, the largest wetlands are found across Canada and in Alaska, Florida, Louisiana, and Minnesota. In other states, wetland **ecosystems** also lie along creeks or near beaches.

There are two main types of wetlands: **coastal** and **inland**. Coastal wetlands lie along oceans, bays, gulfs, and sounds. There, the water is salty or brackish. Inland wetlands form near freshwater rivers, streams, lakes, and ponds. A wetland can cover millions of acres or can be as small as a mud puddle

Salt marshes like this one in Maine are popular breeding grounds for wading birds, frogs, and insects.

that doesn't dry up. Water covering land creates a wetland **environment**.

Common coastal wetlands include swamps, marshes, and tidal flats. Tidal flats, or salt marshes, lie near oceans, seas, and bays. The water level covering the land rises and falls with the tides. Salt-marsh grasses and reeds grow as tall as ten-year-old children do, while bushes lie close to the ground. Marsh trees are few and are usually short and stubby. People who move through salt marshes surprise sea birds that are searching the shallow water for fish, clams, and crabs. Small islands of higher ground in this ecosystem provide dry nesting spots for ducks, geese, gulls, and terns. Here, foxes, rats, and weasels keep an eye out for toads, snakes, and lizards.

CITES

The Convention on International Trade in **Endangered Species** (CITES) of Wild Fauna and Flora is a worldwide treaty that protects endangered plants and animals and prevents animal products from being sold. Some wetland animals, such as alligators and pelicans, became endangered because people sold their skins or feathers.

Freshwater Wetlands

A typical freshwater wetland might be a prairie pothole or a lake. A prairie pothole is a small depression in the land that collects water from melted snow or rain. Potholes are temporary wetlands. The collected water dries up by late summer. In the Great Plains of North America, low-lying basins collect water and create wetland **habitats**. These wetlands are essential for **migrating** birds, such as ducks, geese, and cranes. Between their summer nesting sites in Canada and their winter homes in the South, these birds need wetland areas where they can stop for food and rest.

One of the oddest wetlands is the playa, found in southwestern deserts. Playas are wet only during the rainy season.

Prairie potholes provide a welcome rest stop for ducks, geese, and cranes migrating for breeding.

However, during that time, a playa supports an ecosystem of its own. Tiny fish and reptile eggs, plant seeds, and insect eggs lie in the dry mud of a playa, waiting for the rains to come. When a few inches of water are added, the seeds sprout and the eggs hatch. A new community that may only last a few days or weeks thrives.

What Happened?

Today, less than half of the natural wetlands of North America still exist. Humans have destroyed the natural wetlands by draining the water and using the land for other purposes, polluting the ecosystem.

Water loss destroys the delicate balance of wetland environments. The water might be drained for drinking water or used for watering crops. The area doesn't have to dry up completely for wetlands to be damaged. Sometimes, lowering water levels by just a few inches causes damage. Less water means less space for fish and other water animals and plants to live in. Lowering water levels also reduces the region's ability to clean itself of animal and plant waste.

One of the easiest ways to make new farmland or housing space is by filling in a wetland with dirt. With every dump truck of topsoil that is brought in, some wetland animals and plants lose their homes. Loss of habitat is a problem affecting whooping cranes, wood storks, and spoonbills. The filling in of pools and ponds reduces the number of nesting places available to wading birds. Because these birds eat water-dwelling animals, less water also means that there will be less food available for them.

What Can I Do?

Everything that goes down the drain or on a lawn ends up in the water system. Help stop wetland pollution.

- Keep pet waste, leaves, and trash out of street gutters and storm drains.
- Use only **biodegradable** toilet paper, cleaners, soaps, and pesticides.

Dredging soil from a wetland to build a marina causes as much damage to the ecosystem as filling it in with dirt. Both changes alter water flow, increase pollution, and destroy natural habitats.

Pollution is another serious problem. Pollution can be direct, as in dumping garbage into rivers. It can also be indirect, as happens with **fertilizer** or **pesticides**. Chemicals seep into the soil and flow along the **water table** toward rivers and seas. The pesticide sprayed on a rosebush today will be in a wetland water system within a month or two.

The Everglades

The Everglades is a 4,000-square-mile (10,360-square-kilometer) swamp in Florida. The swamp is more than three times the size of Rhode Island. At the southern end of the swamp lies the Everglades National Park. It covers about 2,185 square miles (5,569 square km) of land. The area is a mixture of swamp, forest, and tall saw grass. Much of the Everglades is covered by a river that is 50-miles (80-kilometers) wide and only 6 inches (15 centimeters) deep in most places. Alligators, brown pelicans, and manatees are among the animal survivors in this wetland ecosystem.

Saving the Wetlands

The **Environmental Protection Agency** (EPA) supports a "Five-Star **Restoration** Program" for wetlands. Part of the plan is to restore 100,000 acres (40,469 hectares) of wetlands yearly between 1999 and 2005. The plan also gives money to environmental groups, states, schools, and tribal groups to support local wetlands programs.

These programs reduce pollution, control water **drainage** and runoff, and protect the wetland habitat. National parks and preserves save some wetland areas, yet others are still being drained for farming or housing. For example, Louisiana's coastal wetlands disappear at a rate of about 35 square miles (91 square kilometers) each year. That is larger than New York's Manhattan Island. The loss of Louisiana's wetlands is important because 95 percent of the fish and marine animals living in the Gulf of Mexico start life or live in this region. Fisheries in the United States take about one-third of their annual catch from the Gulf of Mexico. Loss of the wetlands directly reduces the number of fish and marine animals in the gulf. Saving and restoring the coastal wetlands in Louisiana would cost about $14 billion. Continued loss of this resource may result in a $100 billion loss in jobs and food resources.

Saving the wetlands also means saving the unique animal **species** of wetlands habitats. These watery ecosystems teem with animal life—from tiny gnats to huge alligators. While many wetland creatures teeter on the brink of **extinction**, **conservation** efforts have turned some near tragedies into success stories. North American survival stories include alligators, brown pelicans, manatees, and whooping cranes. Across the Pacific Ocean lives another wetland survivor—the Japanese crane. Once believed extinct, this elegant bird has made a remarkable comeback.

A scientist measures pollution levels in an Arctic lake. Polluted water affects the surrounding wetlands and poisons wildlife in the area.

A young alligator strolls across the University of Florida's Gainesville campus.

Chapter Two

Keystone Species: Alligators

A young bull alligator ambles across the grass near Lake Alice in central Florida. He and his kind have struggled to find safe habitats in Florida wetlands. The campus of University of Florida is a remarkable exception in this area.

In Lake Alice, the alligator is free to sleep in the sun without fear of **poachers**. There are no enemies fighting him for space. Food is plentiful—birds, small mammals, and fish thrive in the area.

There are humans all around Lake Alice, yet no one bothers the bull. He lives safely in the heart of the University of Florida's Gainesville campus. To date, the school hasn't lost a single student to an alligator nor an alligator to a student.

Threats to Survival

Unfortunately, alligators and people haven't always been able to live in the same space. Alligators like warm, wet habitats where they can sun themselves on land and slip quickly into the water when a possible meal comes along. Warm, sunny climates also attract people, and people haven't been too happy about sharing the land with these animals.

Alligators, like all reptiles, are cold-blooded animals. They warm up their body temperatures by sunning themselves.

Alligator Fast Facts

Description:	Long, lizardlike shape with a thick body and long tail; tough skin, sharp teeth, and dull olive or gray coloring
Male:	Bull
Female:	Female
Offspring:	Hatchling, 20 to 60 eggs per nest
Groups:	Hatchlings form a pod, a group of many young living together; adults live alone
Life Span:	30 to 65 years
Size:	**Males:** 10 to 12 feet (3.4 to 3.7 m); 450 to 500 pounds (204 to 249 kilograms) **Females:** 8 to 10 feet (2.4 to 3 m); 150 to 170 pounds (68 to 77 kg)
Food:	Fish, snakes, frogs, turtles, small mammals, and birds
Range/Habitat:	Southeastern United States, including Texas, Louisiana, Mississippi, Alabama, Georgia, Florida, North Carolina, and South Carolina

Alligator relatives have existed for more than 150 million years. Large and powerful, they have few natural enemies, with the exception of humans. In the 1770s, there were so many alligators in Florida's Saint Johns River that naturalist William Bartram said, "It would have been easy to have walked across the river on their heads." Yet, within 350 years of Europeans arriving in North America, the American alligator teetered on the brink of extinction.

There were three main threats to alligators, and all of them came from humans. First, alligator habitats were taken over for farming and housing. Communities also drained wetlands for drinking water or for much-needed freshwater that was used to irrigate crops. However, the greatest threat to alligators was the fashion industry.

In the 1950s, alligators were more likely to be turned into purses or shoes than to live to old age. Hunters grew rich selling alligator skins. Laws to protect the alligator were ignored by poachers, who were too greedy to stop the killing. In 1967, alligators were placed on the endangered species list and protected by federal law.

Saving the Gator

The American alligator's success story is the result of careful planning. The conservation program had to reduce the amount of illegal hunting, allow wild alligators to grow and thrive, and keep a close eye on alligator **populations**.

Protecting wild alligators started with farming—only this time it was alligator farming. People interested in starting alligator farms received permits to collect eggs, hatch them, and raise the hatchlings into adults. Once farms were able to meet the demand for alligator skins and meat, there was no longer enough money to be gained from poaching to make the risks worthwhile for the illegal hunters.

Since the 1970s, wildlife officials have kept a count of alligators in their natural habitats. Once alligators were no longer hunted in large numbers, the wild population bounced back dramatically. The rebound was, in fact, too successful.

There reached a point when there were simply too many alligators. This meant less food and fewer nesting areas for the alligators and the increased likelihood that disease could hurt the population.

Alligator farms separate hatchlings, juveniles, and adults. Otherwise, the adult males might eat the young reptiles.

Once the wild population grew larger than the number the natural habitats could handle, a means of controlling those numbers needed to be found. Oddly enough, the answer was hunting–the original cause of the alligator problem. This time, however, the hunting was carefully controlled. A hunter

Keystone Species

Alligators are a **keystone species** in the southeastern wetlands because they change the habitat around them. As wetland water levels drop, alligators thrash around in muddy water, deepening the pools in which they live. During a drought, alligator pools provide water for many species, such as raccoons and wading birds. Alligator trails and nesting sites also shape the land around them, making paths and refuges for other wetland creatures.

bought a permit that allowed the hunting of a specific number of alligators, usually of a specific size. This kept young alligators safe and allowed the population to be reduced slowly.

Today, state and federal parks create safe habitats for wild alligators. Some of the money used to manage state and federal alligator-protection programs comes from permits sold to people who want to farm or hunt alligators. In this way, the hunt supports the survival of the species.

Gentle Jaws

A female alligator's jaws are strong enough to break a human leg with one bite, yet she carries her hatchlings in her mouth without leaving so much as a scratch on their tiny bodies.

Supporting Survival

As alligators lose their natural habitats, they look for new homes. Able to adapt to most wet environments, alligators are often found in drainage ditches, golf-course water hazards, public fountains, and even swimming pools. State fish and wildlife departments are often called on to remove alligators from populated areas. This happens so often that Florida officials remove four thousand alligators annually from such situations.

Going from near extinction to survival took alligators about thirty years. In 1987, the U.S. Fish and Wildlife Service announced that the American alligator population had recovered. At that time, the species was removed from the endangered species list, but it is still listed as threatened. Alligators are protected under the threatened status because they look very much like **caimans** and crocodiles—both still endangered and struggling to survive. Their similar appearance might cause hunters to mistake an alligator for a crocodile or a caiman.

By 2001, the population count showed that alligators were a thriving species. Among the states in which alligators live, Florida and Louisiana report large alligator populations. In the Everglades alone, there are more than one million alligators—wild, free, and surviving.

The Everglades, a haven for alligators, suffers from reduced water supply and heavy pollution. Cleaning up the Everglades will ensure a safe future for the alligator population.

A pelican skims just above the water's surface looking for schools of small fish.

Chapter Three

Brown Pelicans

A brown pelican glides just above the water's surface. Below, a school of silvery minnows shimmers beneath the quiet bay water. The pelican dives and fills its bill with saltwater and fish.

An old saying claims that pelicans' beaks can hold more than their bellies can. That is true! A pelican's bill and pouch hold up to 3 gallons (11 liters) of water that squeeze out of the beak's sides, leaving only the fish behind.

The pelican flies to its nest with a beak filled with food. It is a good year for the baby pelicans. All three eggs hatched, producing healthy chicks with endless appetites. Papa pelican opens his bill wide as his offspring feast on their minnow supper.

Threats to Survival

Pesticides sprayed on crops seep into the soil and travel along the water table. They change nature's balance in every ecosystem by killing insects, small mammals, and birds.

There have been many threats to the brown pelican's survival. The pelican has few natural enemies, although hurricanes and floods damage many nests each year. As is true for many species, the real threats come from people.

In the late 1800s, pelican feathers were a popular decoration on women's hats. Unfortunately, slaughtering pelicans

was the most common way to get the feathers. This practice was just the beginning of problems for the pelican.

Just after World War I (1914–1918), pelicans were blamed for eating fish that would otherwise be caught by **commercial** fishers, so fishers killed thousands of pelican to protect their catches. Later, scientists proved that pelicans did not feed on the same fish that humans ate, but that information came too late for many pelican flocks.

The pelican population also decreased because of oil spills, the loss of wetland habitats, and getting caught in fishing lines. However, the most deadly force acting against the pelican started doing damage in the 1940s. It was called DDT, a pesticide widely used on farms, on lawns, and in parks to kill harmful insects. DDT weakened the pelicans' eggs and made it difficult for them to produce healthy young birds.

DDT seeped into the soil and into the water table. Worms, plants, and insects that carried DDT were eaten by fish, which, in turn, were eaten by pelicans. Soon, the amount of DDT in many pelicans' bodies was so great that it prevented them from laying strong eggs. DDT made the eggshells too

Rachel Carson and DDT

In 1962, Rachel Carson published *Silent Spring*, a book about the effect of chemicals, fertilizers, and pesticides on the environment. She pointed out to her readers that something that kills a worm or plant would most likely sicken or kill whatever animal eats the worm or plant. Carson is often called "the mother of the environmental movement" because of her efforts to stop chemical pollution.

thin to bear a parent bird's weight. When eggs are cracked or crushed, the chicks inside die. Many of the chicks that did hatch were sickly and died before reaching adulthood. DDT was slowly destroying the pelican population.

Today, wetland and water-table pollution still threatens the brown pelican. The EPA reports that pesticide sales in 1997 topped $11.9 billion. Oil spills and pollution in rivers that feed wetlands also add to the dangers for the brown pelican.

Theodore Roosevelt and Pelican Island

Efforts to save the brown pelican began in 1903, when Theodore Roosevelt named Pelican Island, Florida, as the first national wildlife refuge. Pelican Island protected nesting pelicans from egg hunters and feather collectors.

In 1903, Pelican Island National Wildlife Refuge became the first federally run preserve in the United States.

Brown Pelican Fast Facts

Description:	Large gray-brown bird, about the size of a small turkey, with a white head and neck; exceptionally large, hanging bill and pouch for fishing
Male and Female:	No specific name
Offspring:	Chicks, in clutches of 2 to 3 eggs
Groups:	Adults are a flock; groups of chicks are called pods
Life Span:	15 to 30 years in the wild, up to 40 years in a zoo
Size:	Height is 42 to 54 inches (1.1 to 1.4 m); weight is 8 to 10 pounds (about 4 kg)
Wingspan:	6 to 7 feet (about 2 m)
Food:	Fish, squid, and shrimp
Range/Habitat:	Nests are built in trees or in marshes; range from Virginia to Brazil on the East Coast, California to Chile on the West Coast, and the Caribbean and the Gulf of Mexico

The Migratory Bird Treaty Act of 1918 soon followed. This law was passed in an effort to protect birds such as pelicans, whooping cranes, and trumpeter swans from overhunting. The law made it illegal to hunt, capture, kill, collect, possess, or transport birds, eggs, or nests. The Migratory Bird Treaty Act is still in effect. Over the years, lawmakers have changed and added to the law. Today, a fine for breaking the law can be as much as $15,000.

In the early 1970s, two major events occurred that further helped the brown pelican. The first was the listing of brown pelicans as an endangered species in 1970. From coastal

California, around the Gulf of Mexico, and along the East Coast from Florida to Virginia, endangered-species protection laws guarded pelicans and their nesting sites.

The second important ruling came from the EPA. That group banned the use of DDT and limited the use of other types of pesticides. The pelicans had a chance, but only time would tell if that chance would be enough to save the species.

Less than fifty years ago, pelicans no longer lived in Louisiana. These birds, nesting on North Island, belong to pelican flocks reintroduced to Louisiana's barrier islands.

Recovery Plans for Pelicans

Of the many birds affected by the use of DDT, pelicans were the first to recover. By 1985, the pelican population along the Atlantic coast of North America was large enough for brown

pelicans to be taken off the endangered species list for that region.

In 1968, Louisiana no longer had a wild pelican population. A joint program between Louisiana and Florida wildlife groups began restoring pelican flocks to Louisiana wetlands. Within three years, Louisiana officials found nesting pairs on North Island and Queen Bess/Camp Island. The program has been a moderate success, with the pelican population increasing annually. Another Louisiana program uses soil cleared from the Mississippi River delta to make new barrier islands in the Gulf of Mexico for pelican nesting. So far, the human-made islands have become popular nesting spots for brown pelicans.

Regular counts of nesting pairs, eggs, and chicks help scientists keep track of the brown pelican population. Efforts to protect the water table and pelican nesting sites have saved the brown pelican from extinction.

The International Crane Foundation is one of six facilities that have captive whooping crane populations. Chicks born there may be raised to return to the wild.

Chapter Four

Whooping Cranes Rebound

In truth, there were never all that many whooping cranes. At most, about five thousand whooping cranes may have lived at one time before Europeans came to North America. By the 1800s, hunting, disease, human settlements, and severe weather had reduced the number of cranes to 1,500 birds.

By 1950, the total number of whooping cranes—often called whoopers—in the world was only sixteen. Saving the whooping cranes from extinction demanded desperate measures. Scientists found that while whoopers often laid two or three eggs, they only raised one chick. It might be that whooping cranes can't collect enough food to feed more than one chick. It might also be that some instinct limits their parenting to just one chick. Whatever the reason, a way had to be found to hatch every egg and raise every chick to adulthood.

Sandhill cranes make excellent foster parents for whooping crane chicks. Unfortunately, once the whooper chicks mature they do not mate.

In 1975, scientists borrowed the idea of **fostering** from **falconry**, the sport of raising birds of prey for hunting. Falconers took eggs from one nest and had the chicks hatched and raised by other birds. Would fostering work with whooping cranes?

The extra whooper eggs were removed from every nest that held more than one. Those eggs were placed in sandhill crane nests and hatched successfully. Wild sandhill crane parents raised the whooper chicks as part of their flock. They taught them how to find food, migrate for the winter, and survive in the wild. By 1988, there were twenty whoopers living among the Rocky Mountain sandhill crane flock. Yet, the program was a dismal failure. While the young whoopers thrived, they simply wouldn't mate. No new whoopers were ever born to the sandhill-raised whoopers. Today, only two of the twenty Rocky Mountain whoopers survive.

Whooping Crane Census

This chart shows the progress made in saving the whooping cranes.

Year	Total Population
1860–1870	1,330–1,400
1941	22
1950	16
1960	36
1997	320
2000	388
2003	422

What Whooping Cranes Need to Survive

As of 2003, there were just three hundred wild whoopers. One migrating flock winters in Aransas, Texas, and spends summers in Wood Buffalo Park in Canada. That flock has about 185 birds. A smaller eastern migrating flock has about twenty birds. There are also another 125 whoopers living in **captivity** in zoos and research centers. A non-migrating flock in Florida has around ninety birds.

Whooping cranes need winter and summer homes, as well as rest stops between each home. Migrating whoopers travel more than 2,500 miles (4,000 km) twice a year between Aransas National Wildlife Refuge in Texas and Wood Buffalo National Park in Canada's Northern Territory. Although Aransas and Wood Buffalo provide enough space for whoopers to live, wetland areas where migrating birds rest and feed

In 1993, a flock of whoopers returned to a wild environment in Kissimmee Prairie, Florida. This flock remains in Florida all year, reducing the potential for deaths from migration.

An eastern migratory whooping crane flock spends summers in Necedah National Wildlife Refuge in Wisconsin and winters in the Chassahowitzka Refuge in Florida.

The Gene Problem

Every feature of every species—height, wing shape, and body shape, for example—is carried by **genes**. The genes from all the animals in one species is called a gene pool. Human beings have a huge gene pool; whooping cranes do not.

A chick gets its genes from both parents, and the stronger features are usually passed on. In a flock of five hundred birds, perhaps 20 percent may carry a gene for disease resistance. If a disease strikes the herd, the 20 percent, or one hundred birds, with resistance to the disease will survive. That is enough birds to rebuild the flock. However, if there are only sixteen birds in the whole flock, 20 percent equals three birds. If disease wipes out thirteen birds, the remaining three will not be able to make the flock survive.

are disappearing as people drain the land. When humans fill in and plow over wetlands between Aransas and Wood Buffalo, for example, they eliminate places for the whoopers to rest during their migration. This leads to exhaustion and hunger, killing whoopers during their migration.

Hunting, power-line accidents, and natural disasters take their toll. With such a small wild population, one whooper being shot by a hunter is a disaster. Yearly, power-line collisions kill a few more birds. A hurricane or hailstorm can be a tragedy. In 1940, a hurricane killed six of the thirteen-bird Louisiana flock—more than 45 percent of all Louisiana whoopers.

Wood Buffalo Preserve

In 1937, the United States government set aside Aransas National Wildlife Refuge as a winter home for whooping cranes. However, no one could discover the cranes' summer home in Canada. From 1945 to 1948, a nationwide search for the whooper nesting area took place, with no success.

Whooping cranes were one of the first birds listed as an endangered species.

In 1954, a helicopter pilot scouting Wood Buffalo National Park for fires saw a patch of white in the marsh below. Flying lower, he saw two whoopers and a chick. Eureka! The whooper nesting site had been found. This was just one step in trying to save this endangered bird.

What Next?

Saving the whooping crane meant building up captive and wild flocks. The problem was that there weren't whoopers to start new flocks. A captive breeding center was set up in Patuxent, Maryland. At the Patuxent facility, scientists considered different ways to help preserve the whooping crane.

Current breeding experiments for whoopers are first tried on sandhill cranes, which are plentiful. If an experiment is successful, it is repeated using whooper eggs. If not, valuable whooper eggs have not been wasted. The number of adult whoopers continues to grow slowly.

Whooping Crane Fast Facts

Description:	Tallest North American bird, about the height of an average ten-year-old child, snow-white bodies with black wing tips. Long neck, a dark bill, and a bright red forehead.
Male and Female:	No specific name
Offspring:	Chicks, in clutches of 1 to 3 eggs
Groups:	Adults are a flock
Life Span:	20 to 24 years in the wild, 35 to 40 years in captivity
Size:	Height is about 49 to 60 inches (1.4 to 1.8 m); weight is 14 to 16 pounds (6.4 to 7.3 kg)
Wingspan:	6.5 to 7.5 feet (2 to 2.3 m)
Food:	Insects, minnows, crabs, clams, frogs, mice, berries
Range/Habitat:	Nests in marshes, swamps, wetlands throughout North America; has summer nests in Canada

In 1993, a new flock was set up at Kissimmee Prairie in Florida. These whoopers never learned to migrate. They live in Florida all year. By 2000, the Florida flock numbered seventy-eight whoopers. To the delight of wildlife scientists, one nesting pair in this flock successfully hatched a chick. This was a first! No captive-raised whooper had ever produced a chick in the wild before.

Since efforts to save the whooping crane first began, the numbers of birds has increased slowly but steadily. New habitats and flocks ensure the survival of the whooping crane.

At birth, a manatee calf weighs about 70 pounds (32 kg). Mother and calf communicate by touching with their flippers and by squeaking.

Chapter Five

Hope for the Future: Manatees

A mother manatee, called a cow, slips through a gate on a Florida canal. Sadly, her calf is caught on the other side. Cow and calf call to each other continuously while they are separated. Luckily, this calf is old enough to survive the separation, but this is not always the case. The mother's clicks, squeaks, and squeals assure the calf that he has not been left behind.

Christopher Columbus and his crew thought manatees were mermaids. They obviously didn't get a close look at these mammals.

This mother, like all manatee cows, is devoted to her child. The two touch, kiss, and hold each other, as the mother serves as both parent and best friend to her calf.

The canal gate finally opens after three long hours, and mother and child are reunited. The time apart has been stressful for both manatees because, during their first eighteen to twenty-four months of life, most manatee calves are rarely away from their mothers for more than a few minutes.

Threats to Manatees

Manatees were among the new experiences enjoyed by Christopher Columbus and his crew in 1493. For some reason, they thought the round, gray, homely manatees were mermaids. Columbus himself said, "These mermaids were not quite so handsome as they have been painted." Handsome or ugly, the good-natured manatee has suffered badly from human contact since Columbus's time.

In colonial days, manatees were hunted for their skin, meat, and oil. More recently, human-made locks and canal gates, motorboat propellers, loss of habitat, pollution, and poisonous red tides have killed too many of these gentle mammals.

Manatees are slow swimmers, moving at about 2 miles per hour (3.7 kilometers per hour). They breathe air and often hover just below the surface of the water. They do not move quickly enough to swim through locks or gates that are closing. If they hear approaching boats or barges, manatees are too slow to avoid engine propellers. Many manatees are scarred or killed by locks, gates, and boats each year.

The manatee's habitat is under siege. As more people move to Florida, the demand for freshwater increases. Some communities drain wetlands to get this water. In some areas, wetlands are filled in to provide land for housing. In others, cutting trees and natural shrubs reduces the ground cover that holds down the topsoil. When soil erodes into rivers, it makes the water cloudy and reduces the ability of water plants to survive. In Tampa Bay, for example, about four-fifths of the sea-grass beds have been destroyed. Because sea grass is a primary food supply for manatees, loss of this food source affects their chances of survival.

One deadly event in which humans play no part is a red tide, formed by tiny microorganisms called dinoflagellates. The red tide is poisonous to manatees. The microorganisms are eaten by tiny fish and also cling to sea grass in manatee habitats. When manatees eat red tide–polluted plants, they die. In 1996, 150 Florida manatees died during a red tide. Luckily, red tides are rare.

An Elephant's Cousin?

The manatee's closest relative is not a whale or a dolphin but an elephant!

A red tide in 1996 killed more than 150 Florida manatees. The manatees swallow the poisonous organisms when they eat sea grass.

Manatee Recovery Efforts

Both federal and state laws protect manatees. The federal government passed the Marine Mammal Protection Act of 1972 and the Endangered Species Act of 1973 in an effort to protect all marine mammals. It is a federal crime to bother, harm, hunt, shoot, wound, kill, capture, or collect any endangered species. This includes manatees. A federal recovery plan for manatees was developed by the U.S. Fish and Wildlife Service. This plan includes tagging the manatees, which allows their movements and daily activities to be tracked by radio transmitter. Happily, the plan is proving to be a success.

Because most U.S. manatees live in Florida, the state government supports recovery efforts for these gentle animals. Most manatees live within a thirteen-county area. Within those counties, speed zones for boats, ranger-monitored programs, and local protection plans are in effect. During mating periods, many manatee habitats are off limits to humans.

Captive medical care helps return injured or sick manatees to the wild. Manatees wounded by boats or personal watercraft are scarred by their injuries. A program to catalog these

Manatee Fast Facts

Description:	Large, gray mammals with rough, rubbery skin much like the surface of a tire; little body hair, stiff whiskers, and a paddlelike tail and flippers
Male:	Bull
Female:	Cow
Offspring:	Calf, one born at a time
Life Span:	Ranges from 28 years to 60 years
Size:	Normal length is about 10 to 15 feet (3 to 4.5 m); adult weight ranges from 800 to 1,500 pounds (364 to 682 kg); manatees grow all throughout their lives
Food:	Water plants, such as water hyacinth and hydrilla (up to 100 pounds [45 kg] daily)
Range/Habitat:	Slow-moving rivers and bays, mostly in Florida and the Caribbean

scars helps identify different manatees and their habits.

The Florida Marine Research Institute runs a manatee census or head count twice yearly. In January 2003, the census counted 3,113 manatees, an increase over the population numbers of 1992. While it is impossible to count every single manatee, this is a reasonably accurate number. However, the census only lists the manatees that were visible from the air on census day. Others were not

Biologists and aides examine a manatee for skin damage. They attach a transmitter to the manatee's skin so they can track the mammal in the water.

The Crystal River Refuge

The Crystal River National Wildlife Refuge in Florida is a federal preserve that provides a safe, clean-water environment for manatees. The preserve teaches people about the manatees. This is the only Florida site where people can swim with manatees.

seen and, therefore, not counted. Population enlargement is due to successful protection programs, education, and government and privately sponsored safe havens.

Educating the Public

One way to keep manatees safe is through public education. Most boaters happily follow speed limits and restrictions if they know manatees are in the area. Controlling dumping and polluting in manatee habitats has also become easier once people were made to realize the problems pollution causes.

The Save the Manatee Club sponsors an adoption program through which people "adopt" a manatee and learn about their specific manatee. Adoption fees are used for manatee protection and education programs.

Counting Manatees

Census results for manatees show population growth over the past dozen years.

Year	Total Population
1992	1,856
1996	2,639
2001	3,276
2003	3,113

Manatees have survived in the wild for 45 million years. They can still do so while living beside humans as long as people know what manatees need to survive. Careful management and continued awareness can preserve these affectionate animals for the future.

Boaters must drive slower in manatee areas. Scars from propellers identify many manatees as speedboat accident victims.

Japanese crane chicks learn survival skills by mimicking adult birds.

Chapter Six

The Cranes of Kushiro

Japanese crane chicks believe the first face they see is a parent's face—even if that face is human. This is called **imprinting**. When captive-bred chicks imprint on humans, the chicks will try to learn only from humans. Humans can parent the chicks and teach them to act like cranes, including giving them flying lessons.

Suppose a chick imprinted on a young boy. That chick would naturally copy everything the boy did. To teach the chick to fly, the boy would have to run very quickly with his arms flapping. He would do this repeatedly with the eager chick running and flapping close behind him. Suddenly, the flapping would bear results. The chick would be airborne. Success! One lesson is learned, but there are many more to be learned before a captive-bred chick can survive in the wild.

The End of a Species?

During the late 1800s, people began moving into the region of Hokkaido, Japan, which was once a natural habitat of the Japanese crane. No one worried about losing the crane habitat because Japanese cranes, or tancho cranes, were considered to be extinct. No Japanese cranes had been spotted in the wild for many years. Then, in 1924, a small flock of Japanese cranes was spotted in Kushiro Marsh on the eastern edge of Hokkaido.

The Japanese people wanted to protect the twelve surviving cranes. One way to save the cranes was by saving their habitat in Kushiro Marsh. By the early 1950s, the flock still nested in Kushiro and had almost doubled in size. Then, in 1952, a terrible snowstorm hit the area, piling snowdrifts on the marsh. The snow was so deep that the cranes could not find fish to eat. When the local people saw that the birds were starving, they fed them grain. This was the start of a continuing relationship between Hokkaido's people and the tancho cranes.

A National Legend

A Japanese legend claims that cranes live to be one thousand years old. Because of this, many people consider the tancho cranes to be a symbol of long life and good fortune. For that reason, tancho cranes are often pictured in temples, paintings, and other works of art.

One Man Makes a Difference

For more than thirty years, when heavy rains struck Kushiro, Japan, Ryoji Takahashi collected Japanese crane eggs from nests that were in danger from flooding. Takahashi took on a serious responsibility. Hatching and raising crane chicks is a twenty-four-hour-a-day, seven-days-a-week job.

A chick this size eats hourly.

The eggs were placed in an **incubator** and turned every three hours until they hatched. Takahashi performed this task himself, talking to the soon-to-be chicks while they were still in their eggs. That was the easy part.

Once hatched, the birds had to be fed regularly, which meant providing them with fresh minnows or other small fish every few hours. Takahashi fed the chicks by holding each minnow with a pair of tweezers to copy the position a parent crane would be in during normal chick feeding. He taught the chicks how to be cranes and then reintroduced each one back into the wild.

Ryoji Takahashi almost single-handedly saved the Japanese

Japanese Crane Fast Facts

Description:	Largest of the cranes; snow-white bodies with black heads and necks and bright red foreheads; also called tancho cranes, Japanese cranes, and Manchurian cranes
Male and Female:	No specific name
Offspring:	Chicks, in clutches of 1 to 2 eggs
Groups:	Adults are a flock
Life Span:	25 to 40 years, and up to 70 years in captivity
Size:	Height is about 60 inches (1.5 m); weight is 22 to 31 pounds (10 to 15 kg)
Wingspan:	7 to 8 feet (2.1 to 2.5 m)
Food:	Fish, frogs, and small reptiles
Range/Habitat:	Nests in marshes, swamps, and wetlands in Japan and coastal Asia

crane from extinction. He is called the "Crane Man of Kushiro," and his story has been filmed. Many of his hand-raised chicks and their offspring still live in Kushiro Shitsugen National Park.

Saved From Extinction

In 1987, the Japanese government established Kushiro Shitsugen National Park, Japan's largest wetland, to protect the tancho cranes' habitat. The park is small—only 104 square miles (269 sq km) of land—and very popular with almost one million visitors each year.

Protection and conservation efforts have saved this majestic bird. The flock that once numbered only a dozen now contains nearly seven hundred cranes. This itself creates a new problem, as there is not enough food or nesting areas in Kushiro for this many cranes. The cranes leave the park and look for food on local farms, destroying crops and angering farmers.

Efforts to find another habitat and split the flock have not been successful so far. Land in Japan is precious, which limits the number of possible park locations. Also, few farmers are interested in having a flock of cranes as neighbors. They fear losing their crops to hungry birds.

In fewer than one hundred years, the Japanese crane has gone from presumed extinction to an overly successful flock. The worldwide wild population of Japanese cranes stands at

Male and female Japanese cranes perform an elegant dance when choosing their mates. These cranes become lifelong partners.

about 1,500 birds. Another seven hundred live in zoos. The success of the legendary tancho crane appears to be both a blessing and a burden. The Japanese government and conservation groups will continue to look for new habitats that will support this remarkable species.

Glossary

biodegradable—easily decayed and absorbed by nature

brackish water—mixed fresh and salty water

caiman—a reptile that looks much like an alligator

captivity—a state of being kept in a park, zoo, or nature preserve

coastal—along a sea, gulf, or ocean

commercial—for business or sale

conservation—careful management and use of natural resources, such as forests, lakes, rivers, or wild animals

drainage—flow of water out of an area

ecosystem—the animals, plants, and features of an area

endangered species—animals and plants that are threatened with extinction

environment—the earth, air, and water around a living thing

Environmental Protection Agency—a government group that monitors air and water in the United States, and oversees safe use of chemicals and pesticides

extinction—a situation when every member of a species has died

falconry—the sport of raising falcons and teaching them to hunt

fertilizer—a chemical or natural substance that helps plants grow better

fostering—raising young born to other parents

gene—a unit that carries traits or features that are inherited, such as eye color or height

habitat—the place in which a plant or animal usually lives

hatchling—a baby that comes out of an egg

imprinting—the way a very young animal learns to bond with its own kind

incubator—a heated box used to hatch eggs

inland—not along the coast

keystone species—an animal or plant that has a major effect on the lives of other animals and plants in its natural habitat

migrating—traveling from one area to another

pesticide—a substance that kills insects or other pests, such as rats and mice

poacher—a person who hunts animals illegally

pollution—substances in the air, soil, or water that can be harmful to life

population—the number of people or animals in a group

restoration—putting something back in its original condition

species—a group of animals or plants that are alike

water table—the depth at which the ground is saturated with water

To Find Out More

Books

Carroll, David M. *Swampwalker's Journal: A Wetlands Year.* Boston: Mariner Books, 2001.

Cone, Molly. *Squishy, Misty, Damp & Muddy: The In-Between World of Wetlands.* San Francisco: Sierra Club Books for Children, 1996.

Dudley, Karen, Patricia Miller-Schroeder, and Marie Levine. *Alligators and Crocodiles.* Austin, Texas: Raintree/Steck Vaughn, 1998.

DuTemple, Lesley A. *North American Cranes.* Minneapolis: Carolrhoda Books, Inc., 1999.

Greenaway, Theresa. *Swamp Life*. New York: Dorling Kindersley, 1993.

Walker, Sally M. *Manatees.* Minneapolis: Carolrhoda Books, Inc., 1999.

Videos

Audubon's Animal Adventures: Alligator & Crocodile. (ASIN: 0783110782). New York: Audubon Society, 1997.

Conserving America: Wetlands. (ASIN: 6303316328).

Everglades—Secrets of the Swamp. (ASIN: 0792252365). Washington, D.C.: National Geographic Society, 1997.

Wild Wetlands. (ASIN: 193102121X). Animal Safari, Volume 7.

Organizations and Online Sites

American Bird Conservancy
P. O. Box 249
The Plains, VA 20198
http://www.abcbirds.org
Find out how this organization works to preserve birds and their habitats.

Florida Manatee Research & Education Foundation
12025 North Elkcam Blvd.
Dunnellon, FL 34433
Learn about manatees through this group's educational materials, which are suitable for children.

International Crane Foundation
E-11376 Shady Lane Road
P. O. Box 447
Baraboo, WI 53913
http://www.savingcranes.org
Write to the International Crane Foundation for crane fact sheets, teaching materials, and audio-visual materials.

National Audubon Society
700 Broadway
New York, NY 10003
http://www.audubon.org
Learn about birds and bird watching from the National Audubon Society.

National Wetlands Research Center
http://www.nwrc.nbs.gov
This government-funded group investigates plants and animals found in wetlands.

Patuxent Wildlife Research Center/Whooping Cranes
http://whoopers.usgs.gov
This is a delightful site with the most up-to-date information about whooping cranes. It also features videos of whoopers.

Save the Manatee Club
500 North Maitland Avenue, Suite 210
Maitland, FL 32751
http://www.savethemanatee.org
Take part in the effort of this group to save to manatees and their habitats.

A Note on Sources

We hear so much about animal species on the verge of extinction that I wanted to write about animals on the road to recovery: the survivors. Research for this book series began on the Internet at web sites hosted by U.S. Fish and Wildlife and the American Zoo Association. *National Geographic* and *Smithsonian* magazines featured articles on conservation efforts for many animals.

Many thanks to Dr. Bruce B. Ackerman of the Florida Marine Research Institute, Nancy Higginbotham of the Louisiana Department of Wildlife, and Bill Vermillion of the U.S. Fish and Wildlife Service for their advice on animals in this book.

Children's books helpful in researching this book include *Endangered Wetland Animals* by Dave Taylor and *North American Cranes* by Lesley A. DuTemple. Ask your local librarian to help you find quality research material on any topic.

—Barbara Somervill

Index

Numbers in *italics* indicate illustrations.

About the Author

Barbara Somervill has been fascinated by alligators and manatees since her first trip to Florida as a toddler. Since then, she's visited Florida many times and enjoyed the varied wildlife the state has to offer. The story about the Crane Man of Kushiro interests Barbara because it shows how one person can make a difference in saving an endangered species.

Freelance writing keeps Barbara busy. She writes books for children, video scripts, magazine articles, and textbooks. The strangest subject she has written about was coffins!

Barbara was raised and educated in New York. She's also lived in Toronto, Canada; Canberra, Australia; California; and South Carolina. She is an avid reader and traveler, and she enjoys learning new things every day.